GAME BASED LE~

IN EDUCATION 4.0

[2ND EDITION]

PETER CHEW

HAYDAR AKCA

MASANORI FUKUI

NG KHAR THOE

PCET VENTURES (003368687-P)

Email:peterchew999@hotmail.my

© Peter Chew 2023

Cover Design : Peter Chew

Cover Image: Freepik Premium

Mathematician, Inventor and Biochemist Peter Chew

Peter Chew is Mathematician, Inventor and Biochemist. Global issue analyst, Reviewer for Europe Publisher, Engineering Mathematics Lecturer and President of Research and Development Secondary School (IND) for Kedah State Association [2015-18].

Peter Chew received the Certificate of appreciation from Malaysian Health Minister Datuk Seri Dr. Adam Baba(2021), PSB Singapore. National QC Convention STAR AWARD (2 STAR), 2019 Outstanding Analyst Award from IMRF (International Multidisciplinary Research Foundation), IMFR Inventor Award 2020 , the Best Presentation Award at the 8th

International Conference on Engineering Mathematics and Physics ICEMP 2019 in Ningbo, China , Excellent award (Silver) of the virtual International, Invention, Innovation & Design Competition 2020 (3iDC2020) and Jury in the International Teaching and Learning Invention, Innovation Competition (iTaLiiC2023).

Analytical articles published in local and international media. Author for more than 60 Books , 8 preprint articles published in the World Health Organization (WHO) and 36 article published in the Europe PMC.

Peter Chew also is CEO PCET, Ventures, Malaysia, PCET is a long research associate of IMRF (International Multidisciplinary Research Foundation), Institute of higher Education & Research with its HQ at India and Academic Chapters all over the world, PCET also Conference Partner in CoSMEd2021 by SEAMEO RECSAM.

Peter Chew as 2nd Plenary Speaker the 6th International Multidisciplinary Research Conference with a Mindanao Zonal Assembly on January 14, 2023, at the Immaculate Conception University, Bajada Campus, Davao City.

Keynote Speaker of the 8th International Conference on Computer Engineering and Mathematical Sciences (ICCEMS 2019) , the International Conference on Applications of Physics , Chemistry & Engineering Sciences, ICPCE 2020 , 2nd Global Summit on Public Health and Preventive Medicine (GSPHPM2023) June 19, 2023 and World BIOPOLYMERS & POLYMER CHEMISTRY CONGRESS" 10-11 July 2023 | Online by Drug Delivery,

Special Talk Speaker at the 2019 International Conference on Advances in Mathematics, Statistics and Computer Science, the 100th CONF of the IMRF,2019, Goa , India.

Invite Speaker of the 24th Asian Mathematical Technology Conference (ATCM 2019) Leshan China , the 5^{th}(2020), 6th (2021) and 7th (2022) International Conference on Management, Engineering, Science, Social Sciences and Humanities by Society For Research Development(SRD) and 12th International Conference on Engineering Mathematics and Physics (July 5-7, 2023 in Kuala Lumpur, Malaysia).

Peter Chew is also Program Chair for the 11th International Conference on Engineering Mathematics and Physics (ICEMP 2022, Saint-Étienne, France | July 7-9, 2022) and Program Chair for the 12th International Conference on Engineering Mathematics and Physics (ICEMP 2023, Kuala Lumpur, Malaysia | July 5-7, 2023).

For more information, please get it from this link Orcid: https://orcid.org/0000-0002-5935-3041.

Prof. Haydar Akca

Prof. Haydar Akca graduated in Mathematics and Astronomy from Ege University, Faculty of Science, in 1970. Dr. Akca received Ph. D. in Applied Mathematics from Inonu University, Malatya with collaboration Helsinki University of Technology in 1983.

Since then, he has been teaching in various universities. He becomes Professor in Applied Mathematics at Akdeniz University, Antalya in 1996. He has around 135 technical publications including two monographs.

Prof. Akca's research interest areas; primarily functional differential equations, neural networks, mathematical modelling, control theory, numerical analysis, and wavelet neural networks. He has been organizing serial International Conference on Dynamical Systems and Applications.

Dr. Akca is a member of a number of professional mathematical associations. He is the Editor-in-Chief, and Editorial Board Member of number of International Mathematical Journals.

At present, he is Professor of Applied Mathematics at the Abu Dhabi University, College of Arts and Science Department of Mathematics and Statistics, Abu Dhabi, United Arab Emirates.

Email: haydar.akca@adu.ac.ae , akcahy@yahoo.com

Website: https://sites.google.com/view/prof-haydar-akca/home
https://scholar.google.com/citations?view_op=list_works&hl=en&user=OQIdae4AAAAJ

Associate Professor Masanori Fukui

Assoc. Prof. Masanori Fukui graduated from the Faculty of Science, Himeji Institute of Technology in 2002, the Graduate School of Education, Hyogo University of Teacher Education in 2018, and completed the Graduate School of Engineering, Hiroshima University doctoral course in 2021. (M. A. in Education, Ph.D. in Engineering).

He was a JSPS Special Research Fellow (DC1), a project assistant professor at Hyogo University of Teacher Education until Marth 2021, and has been an associate professor at Tokushima University since April 2021.

Besides, he plans to live in Penang (Malaysia) for a year to conduct the following research. (1) UCSI university involving Dr. Ng and Prof. Dr. Ong (July to September 2023); (2) Wawasan Open University involving Dr. Ng, Assoc. Prof. Dr. Chew Cheng Meng, Assoc. Prof. Dr. Chow and Dr. Tan Saw Fen (September 2023 to February 2024).

Assoc. Prof. Masanori's research interests include learning technology, educational technology, game education, creativity education, programming education, mathematics education, educational psychology, social psychology, and human enhancement using technology, including artificial intelligence.

In the past five years, the total number of refereed papers and refereed international conferences has exceeded 60. Besides, he received 11 academic and educational awards in the past five years, 5 KAKENHI (representative), and 4 Foundation Grants other than KAKENHI (representative).

Adjunct Associate Professor Ng Khar Thoe

Having multidisciplinary interests and a passion for lifelong learning, Associate Professor Ng Khar Thoe has diversified her scope of studies to be an advocate for Science/Social Science education integrating technology-enhanced transdisciplinary approaches. She holds a Ph.D in Education (Open University Malaysia), M.Ed. (Brunel University of West London) and B.Sc.Ed. (Hons.)(University Science Malaysia). Her research interests lie mainly on enhancing essential skills of learners through technology-enhanced science/social science pedagogical approaches including PBA/PBL/IBL. She worked as teacher at primary/secondary levels at schools locally/abroad. She was tutor/trainer/ academic facilitator/specialist at regional training centre; project/thesis supervisor and proposal examiner/Adjunct Assistant Professor (with one doctoral student graduated and another ten in progress) at 3 universities. She has diverse managerial experience as Regional Coordinator/Director,

Acting Head of Science, Editor-In-Chief, Journal Manager and Senior Editor of journals as well as Co-Chair, Publicity Chair and Executive Secretary of Conferences. She was the founder for 'Southeast Asia Regional Capacity-enhancement Hub' (SEARCH) and several hyperlinked project-based programmes including LeSMaT(Borderless) and its offshoot programme Education 4.0 project funded by SEAMEO for Inter-Centre Collaboration (ICC). She was involved in many national/regional/ international programmes as founder/coordinator/judge(1996-2023)/chief judge / reviewer /advisor/ speaker/consultant, also as visiting professor, invited keynote/plenary speaker,

Regional Director of Science across the World (Asia Pacific region). She is life member and Technical Advisor of Society for Research Development (SRD). She published extensively involving as lead writer (>350), editor/chief editor, of which many books/modules/book chapters, also refereed/Scopus, ISI/WoS-indexed high impact publications were cited/referred widely. Her efforts made in these publications also won acclaims/accolades/appreciation from many events at regional/international levels, e.g. ICDE2009, ICDE2013, ISCIIID 2014, 1RCCS2014, 2RCCS2015, iPEINX2016,

ICRDSTHM-17, ICRTSTMSD-18, SEAQIL REGRANT 2017-2019, iCon-MESSSH-20, ICONESS2021, iCon-MESSSH-21, e-STOMATA2022, e-Delphi2022, Minecraft Championship-21-23, e-Sport 2022, iCRI-2022-2023, ICONESS2023, to name a few.

Her current focus/research interests include helping as many youths/ people as possible to achieve their life goals through continuous professional development for inspiring creativity, networking and sustainable living for all. In 2021, the three Minecraft projects she supervised won Top 11th to 20th placing (Bronze). In 2022, the two Minecraft projects that she supervised won No. 4th, No. 8th (Gold,Silver, also the latter team won Champion during e-sport on 18/9), 4 Minecraft projects won Top 11th to 17th placing (Bronze) in 2022 and 1 Minecraft project won 7th placing in 2023.

Her emails: postgradreview@gmail.com, drivng@gmail.com and online CVs are accessible from: https://orcid.org/0000-0002-4462-657X; http://tinyurls.com/cvdrnktupdates; http://bit.ly/cvdrnktupdates & https://myjms.mohe.gov.my/index.php/lsm/about/editorialTeam

Peter Chew Triangle Diagram(preprint) is share at World Health Organization because the purpose of Peter Chew Triangle Diagram is to help teaching mathematics, especially when similar covid-19 problems arise in the future.

https://pesquisa.bvsalud.org/global-literature-on-novel-coronavirus-2019-ncov/resource/en/ppzbmed-10.20944.preprints202106.0221.v1

GAME BASE LEARNING IN EDUCATION 4.0
[2ND EDITION]

TABLE OF CONTENTS

Game Based Learning In Education 4.0
[2nd Edition]

Industry 4.0[1] requires employees who are critical thinkers, innovators, digitally skilled and problem solvers. The problem in the future is not the lack of jobs, but the lack of skills that new jobs will demand. Therefore, Education 4.0 must develop the skills required by the Industry 4.0 labour market, such as problem-solving and digital skills.

Digital Game Based Learning in Education 4.0 can develop the problem-solving and digital skills. Brain training games can also help students develop their working and short-term memory, as well as students processing and problem-solving skills.

The traditional teaching method is that the teacher asks the students to memorise the rules of mathematics, but some students are too lazy to memorize the rules of mathematics, because it is not what they want. If teachers force students to memorise math rules, it may cause students to develop a phobia of mathematics. Through the game-based learning method, because students want to win the game, students will ask the teacher to teach them the rules, which makes it easy for students to remember the rules, because remembering the rules has become what they want.

In addition, students need to concentrate on thinking about the rules while playing the game, which allows them to memorize the math rules for a longer time. This will make teaching and learning mathematics more effective.Playing memory games can help students focus better. Brain training games can also help students develop their working memory and short-term memory, and enhance students' ability to process and solve problems.

Workshop Game Based Learning for memorize maths rule has been held at Conference on Science and Mathematics Education [COSMed 2019] by Southeast Asian Ministers of Education Organization, Regional Centre for Education in Science and Mathematics (SEAMEO RECSAM) . Videos related to this can be found on this link https://youtu.be/yV5soDsi7TU .

2nd Edition update some information and add information of Workshop Game Based Learning for memorize maths rule held at the Center Prai District Education Office, Penang , Malaysia. I would also like to sincerely thank Prof. Haydar Akca , Associate Prof. Masanori Fukui and Adjunct Associate Prof. Ng Khar Thoe for sharing some thoughts in this book.

Peter Chew

Mathematician , Inventor and Biochemist.

Chapter 1: Game Based Learning
In Education 4.0 | 2nd edition |

1.1 WHAT IS EDUCATION 4.0 ? [1]

Education 4.0 is a new experience-based education system that uses digital technologies instead of the rote-based system and responds to the needs of the new world through personalised education. This system, which envisions the training of new generations to meet the needs of Industry 4.0, brings together technology, individuality, and discovery-based learning. It also prepares your children for their **future jobs**.

Education 1.0, Education 2.0, and Education 3.0, and why are they different to Education 4.0 ?

- **Industry 1.0**(1784): Based on water and steam-powered mechanical production equipment.
- **Industry 2.0**(1870): Based on the division of labour and mass production through the use of electrical energy.
- **Industry 3.0**(1969): Based on the use of electronic and information technologies to further automate production.
- **Industry 4.0**(nowadays): Based on the use of cyber-physical systems to lift the boundary between the real and the virtual world.

Industry 4.0 is the door to the modern world, and includes *Cybernetics, Robotics, Big Data, Nanotechnology, Artificial Intelligence, Automation, Global Citizenship, Digital Age* and many more concepts to our lives. With Education 4.0, the concept of education changes completely and some new trends appear that we are not familiar with.

The new education system, built on **success in life and not on exams**, draws attention to the necessity of personalised education. Some novelties that Education 4.0 for Industry 4.0 offers are:

1- Time and space-independent education

Students have the opportunity to learn wherever and whenever they want. Thanks to new interactive learning tools, education is now space and time-independent. While the theoretical dimension is learned outside the classroom, face-to-face practical learning is carried out in the classroom.

Students' need and dependence on buildings surrounded by stone walls, which we call a "class" or "school", are diminishing. Likewise, with the advantage of time independence, a child can undertake their education through e-learning tools and edutainment from their room. Individuals who learn the theoretical part of their education on their own and in a digital environment can transform their knowledge into real-life experiences through practical project-based activities in the classroom.

2- Personalised learning

Students will receive **personalised learning** through special tools adjusted to their capabilities. In this way, a student who has difficulty in understanding what many children can easily absorb will be able to improve at their own pace. Intermediary software will not advance lessons until the student is ready. This advantage of personalised learning means that students who experience learning difficulties at school will feel more supported in overcoming their challenges.

When performance graphs are obtained from this software and examined by teachers, it is possible to understand better the subjects that the **student has a predisposition towards**. Teachers can then easily and effectively determine the subjects in which the student is stronger or weaker.

3- Learning flexibility

Traditional education systems apply the same model for each student. Education 4.0 believes that there is no drawback in trying different paths as long as it reaches the same goal.

From Education 1.0 to Education 3.0, the same curriculum was offered to all students with the same teaching styles. Although more efficient methods were used in Education 3.0, the necessary flexibility could not be provided.

In **Education 4.0**, a flexible global education model is recommended for every student. Teachers can use online data to track and measure the results of their students and then provide personalised guidance based on their specific strengths and weaknesses.

With successful orientation, each child will become more successful in the areas they are predisposed towards and will develop their weaknesses through their own flexible education plan. A successful student in mathematics can complete their development in a weaker verbal field with a personalised and flexible learning plan.

4- Project-based learning (MAKER)

In order to prepare children for the future freelance work model, students need to become familiar with project-based learning and study models. In other words, students have the chance to apply what they have learned on a real project, **instead of writing answers on paper**.

In the field of learning called **Maker**, the individual is transformed into a self-sufficient person by using their talents effectively and in a fun way in many areas, especially technology. Maker culture aims to prepare children for the future via fun.

Through project-based learning, children can improve the below abilities and develop themselves in these areas of great importance throughout their academic career:

- Problem-solving
- Being solution-oriented
- Collaboration and teamwork

- Time management

An example of Maker work is programming-based robotic designs that students develop to perform a specific task.

5- Data interpretation

Mathematics will keep its place in our lives in the future, but this time robots will do these operations instead of humans. **The task for people will be to draw insights based on the released data**. The world is developing technologically every day. Information technologies are the biggest opportunity of our future. When future graduates leave university, they may not know their professions but they will know technology very well and be able to respond to the global needs of Industry 4.0.

It is necessary for people to learn competencies such as setting up, managing, developing, collecting, processing, and interpreting data. As one of the important requirements of Education 4.0, students should be able to recognise trends in data and develop recommendations based on the data. As a result, students should learn to approach standard data from an unusual perspective.

6- Not a single exam, continuous improvement!

In the current system, students are subjected to a question-and-answer exam. According to many educators, this system is only successful in the short-term. After the exam, students forget the memorised information very quickly.In Education 4.0, the focus is on evaluation instead of exams. Students should use their acquired knowledge as soon as they begin their professional life. The evaluation result will be based on the **entire education period instead of one exam**.

Likewise, students are expected to produce continuous Maker activities and put what they have learned into practice. Children learning to code can develop a calculator or a game that they can use in their daily lives, instead of just mastering theoretical knowledge. These projects will contribute to transforming theoretical knowledge into practical experience and storing it in **long-term memory**.

7- Curriculum with student participation

In Education 4.0, students will be involved in the creation of curricula. This is because maintaining a contemporary, up-to-date and useful curriculum will be important to professionals, as well as students.

Students' critical input on the content of their courses will help create an all-inclusive study program that matches their interests. This means that in the future, learning curricula will be prepared by teachers and students together. Currently, learning curricula is only prepared by teachers and contains a significant amount of information that is inapplicable to real-world scenarios,

8- Guidance-oriented

It is believed that in 20 years, students' learning process will have a more independent form. In this form, mentoring will gain importance in order for students to improve their education in a healthy way. Students can achieve the highest level of academic performance under the mentorship of teachers. According to long-term estimates, after 20 years, teachers will play a key role in education as mentors rather than simply distributors knowledge.

John Dewey

"If we educate today's students as if they were living yesterday, we'd steal their tomorrow."

Game Based Learning

According to current research, the knowledge that people discover and experience themselves is more long-term compared to memorized values.

Playing a simple ball game in gamified <u>educational apps for kids</u> can make them learn the laws of gravity and angles without realizing it.

Starting with the example above, we can ask ourselves the following question:

If a computer game is unwittingly able to teach these concepts to children **in an experiential way**, why not apply it to their education too?

This question leads to the emergence of the Education 4.0 approach, as well as the question of how to develop qualified professionals that the world of Industry 4.0 needs.

1.2 An Integration of Game-based Learning in a Classroom: An Overview.[2]

Games are increasingly becoming common in learning environments, and to match the requirements of developing a course as a game, a variety of technologies have been developed. Besides education, game-based learning has been popular in other settings, including professional training and social media.

By introducing gaming elements as a training method, game-based learning platforms will boost students' engagement, motivation, and productivity. Game-based learning is more than just making games for students to play on the surface; it is also about establishing learning activities that gradually teach subjects and lead users to achieve goals.

This paper reviews the two research elements: the instruments used and disciplines done by 21 researchers on the game-based learning approach.

The findings show that most studies have remarkable results in developing the game-based learning method.

Generally, game-based learning in education strikes the perfect balance between subject matter and gameplay and the students' capacity to retain and apply that knowledge in real life. Nevertheless, more research in game-based learning qualities and elements is required to provide extensive knowledge on the adaptability of current technological systems.

According to Saad et al. Technology has taken over worldwide.[3] which makes most people, including educators, need to accommodate the current ways of teaching to satisfy the students' interest in learning.

This method can be used as a game-based approach in English teaching. It is one of the required instructional methods to stimulate students' learning interests and improve teaching effectiveness.[4]

Hence, this study aims to identify the level of participants, examine the methodological approaches used by researchers, and observe the disciplines covered in the selected articles on game-based learning from 2016 to 2021.

This article will also explore the basic foundation and characteristics of game-based learning, the overview of game-based learning from previous research, the implementation of game-based education to the students and teachers, and the conclusion and recommendations for future research.

Theoretical Framework

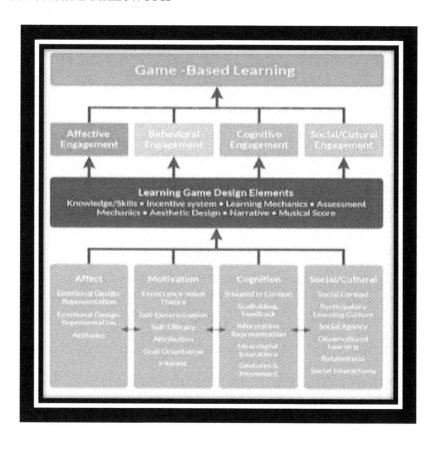

1.3 How to build better
memory training games.[5]

In particular, the field of PL has identified numerous mechanisms (including attention, reinforcement, multisensory facilitation and multi-stimulus training) that promote brain plasticity.

Also, computer science has made great progress in the scientific approach to game design that can be used to create engaging environments for learning. We suggest that approaches integrating knowledge across these fields may lead to a more effective working memory(WM) interventions and better reflect real world conditions.

Video Games

There are many examples of off-the-shelf video games leading to substantial improvements in a variety of perceptual and cognitive abilities. For example,[6] found that training novices for 10 h on an action video game improved performance on enumeration, useful field of view, and attentional blink tasks when compared to participants trained with a non-action video game.[7] found that playing a real-time strategy game improved executive control as measured by task switching, visual short-term memory, and reasoning in older adults.

Another study,[8] showed that an off-the-shelf video game (Portal 2) led to substantial improvements on measures of problem solving, spatial skills, and persistence (in fact, even more so than training with the popular brain training games of Lumosity). Furthermore, dyslexic children improved reading speed and attentional abilities after playing an action video game.[9]

Finally ,[10] employed several computer games targeting executive control, and found improvements in attention, inhibitory control, and planning, which also translated to school performance. Together these studies suggest that video games include important attributes that contribute to learning.

Given the attractive motivational features of video games, recent research in cognitive science is increasingly moving towards adding game-like elements to their assessments. However, without proper design these can impair task performance, and even weaken test quality and learning[11,12]. We suggest a better approach is to create training software that will dovetail, and/or implement non-competing concepts from game design that support learning. The video-game field is maturing, proper design rules and constraints are becoming more refined and the practices of coordinated design are becoming better understood and documented.[13] For example, in order to optimally engage players games must establish clear goals and allow players to realize those goals through meaningful actions.[14] Successful game design has critical aspects that make software engaging, including its mechanics, interaction, visual/sensory experience, and progression.[15]

Many game design criteria mirror components found to improve learning from the PL literature, and literature on deliberate practice.[16] For example, consistent reinforcement to training stimuli,[17] maps directly to consistent player feedback, a key part of *player-centric interface design* .[18] Adams says (of players) "most critically, they need information about whether their efforts are succeeding or failing, taking them closer to victory or closer to defeat".

Likewise, motivating tasks[19] and ensuring subjects are confident of their performance tasks[20] are consistent with good game-design principles such as establishing clear goals goals[21] and balancing games challenges (related to the adaptive approaches used in PL and WM training) to match player performance.[18] Applying video-game techniques purposefully into WM training can inject the cognitive benefits found from off-the-shelf video games into principled cognitive training, while also being fun to play.

Integrating Learning and Gaming Principles

Two relevant lines of research have made significant breakthroughs in brain training:

(1) Studying incidental benefits of off-the-shelf video games;

(2) Transforming standard cognitive tasks into training tasks.

We suggest that the most success will come from integrating knowledge of memory systems with that of brain plasticity and modern game-design principles.

As a first attempt to implement this approach, we created a prototype game that incorporates mechanics of the n-back into an engaging 3D space-themed game (see Figure below).

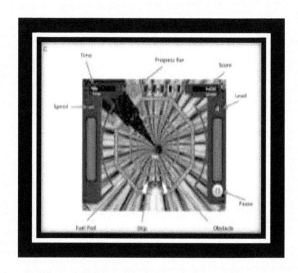

Typically, the n-back task is very basic, e.g., selecting matches from a grid or a picture series. In contrast, our prototype is a space-themed "collection" game with navigation challenges and obstacles, multi-layered progression through levels, and rich, thematic visual and sound effects.

The n-back task is integrated into the game mechanics, where players select the "right" fuel cells while avoiding decoys. Levels are designed to get progressively harder through increasing cognitive challenge (n-level) and other game challenges (such as obstacles). While the game is more difficult and attention is spread over more elements than the conventional n-back, participant's control over their environment is anticipated to increase their engagement with the game.

The game also incorporates principles from PL, where participants are trained on multisensory (auditory and visual) features, where sounds and visuals are designed to facilitate each other, and where attention and reinforcement are carefully sculpted to lead to the best learning.

While much work is still required to maximize the game's efficacy, e.g., by incorporating a broader stimulus set, adding other memory tasks, and creating an even more compelling game framework, we put it forward as a first example of how to build such an integrative game. Initial piloting with our prototype indicates participants are engaged in the game and improve performance (n-level) across training sessions. However, more research is needed to make firm conclusions regarding its transfer potential.

In summary, we suggest that more integrative approaches will lead to better learning outcomes. We suggest that the general mechanisms that promote PL are shared across brain regions and will also promote WM. Furthermore, there is enough known about the aspects of conventional video games that lead to positive learning outcomes that these principles can be applied to achieve more effective WM training.

Additionally, there are other principles, that were beyond the scope of the present review, such as deliberate practice,[22] and many aspects of healthy lifestyles,[23,24] that also promote cognitive fitness. Integrating these approaches with good design could lead to a more comprehensive impact on WM function that might ultimately transfer to real-world conditions.

1.4 The Best Mental Games and Exercises to Improve Memory.[25] *Jun* 14, 2021

When we go to the doctor, we can forget up to 80% of what we're told by the time we get to the parking lot. There are a few things that can cause this kind of short-term memory loss. For example, a shocking diagnosis can make someone unable to concentrate on what the doctor was saying, and therefore unable to absorb information (focus and memory are dependent on each other). Or, if you are given multi-step instructions, you may only remember a few of the items, perhaps in the wrong order—a natural result of new information overload. The root causes of a less-than-stellar memory are many. (Learn more about the four things most likely affecting your brain function.)

There are ways to improve memory and focus through brain training games and exercises, as well as reinforcing good habits. Before we jump into the best mental games and exercises to improve memory, it's important to understand *how* and *why* these memorization games work.

Games That Improve Memory on the Total Brain App

Total Brain's memory-oriented games leverage proven techniques to improve memory:

Faces and Names

If you have trouble memorizing and recalling essential information about new people you meet, Faces and Names can help. The game "introduces" a series of new people in rounds, including their names and occupations, to better develop your ability to make associations for improved recall. Associating new information with an unrelated item is a useful hack for quickly memorizing something. In this example, someone's plane leaves at 2 p.m.; in order to remember the time, they picture the plane's two wings, saying, "My two-winged plane leaves at 2 p.m." This is your associative memory at work.

Memory Sequence

To improve short-term memorization and retention, try playing Memory Sequence. In this game to improve memory, players are faced with a circle divided into four colors (think of the old-school game Simon).

Colors light up, corresponding with numbers and sounds. Players must mimic the order correctly to move up to the next level, starting with a series of four and becoming longer from there.

When it comes to remembering the order of things, it's easier to remember the first and last items in a sequence. Memory Sequence trains your brain to be better at memorizing the "middle content" by repeatedly asking it to retain the entire order to win. Strengthening this part of the memory can help with everyday tasks, like remembering grocery lists. And here's another trick for memorizing the order of things: use acronyms (the first letter of each fact or items into a vivid or memorable word, phrase, or story).

Think of the memorization tricks we were all taught in grade school: "PEMDAS" ("Please Excuse My Dear Aunt Sally") to remember the order of solving an equation, or the imaginary character "ROY G. BIV" for the order of rainbow colors.

Memory Sequence also helps with the memorization trick of chunking, in which we organize the information we want to remember into groups of three. You can enhance chunking by using rhymes, which can more easily build links between pieces of information.

Memory Maze

We mentioned that memory is part of the cognitive function of the brain, which also involves making decisions—so it makes sense that you can improve your memory by training with Memory Maze, which helps players become experts at problem solving in the moment. In this game, players have to remember a path from one end to the other, laid out in a grid of dots. After seeing the path appear for a few seconds, players must repeat the same steps to get to the end. The better you get at remembering the paths, the more complex the routes get.

This taps into our brain's love of patterns. The brain is a pattern-generating system, and seeks them out. Once a pattern is identified, it becomes easier to remember multiple details under an umbrella of information.

Think Focus

We also mentioned earlier in the post that it's impossible to remember something we never really absorbed or processed— which happens a lot thanks to all the modern distractions in our daily lives.

In a time where we can get messages instantly from anyone on multiple platforms, many of us have found help in training our brains to focus adequately and dismiss distractions. That's where Think Focus can help improve focus and attention for stronger information retention and memory formation.

In this game, players have to balance a digital basketball that's spinning on the screen, elevated above the court. The goal is to keep your finger on the ball to keep it above ground. Throughout this timed game, the basketball makes very slight movements that require strong attention and quick reactions, which will move players up to the next level.

To really improve memory, games should be played several times a week. In a month, take the Total Brain assessment again to measure your improvement.

1.5 Video games can change your brain.[26]

Studies investigating how playing video games can affect the brain have shown that they can cause changes in many brain regions

Scientists have collected and summarized studies looking at how video games can shape our brains and behaviour. Research to date suggests that playing video games can change the brain regions responsible for attention and visuospatial skills and make them more efficient. The researchers also looked at studies exploring brain regions associated with the reward system, and how these are related to video game addiction.

The studies show that playing video games can change how our brains perform, and even their structure. For example, playing video games affects our attention, and some studies found that gamers show improvements in several types of attention, such as sustained attention or selective attention.

The brain regions involved in attention are also more efficient in gamers and require less activation to sustain attention on demanding tasks.

There is also evidence that video games can increase the size and efficiency of brain regions related to visuospatial skills. For example, the right hippocampus was enlarged in both long-term gamers and volunteers following a video game training program.

Video games can also be addictive, and this kind of addiction is called "Internet gaming disorder." Researchers have found functional and structural changes in the neural reward system in gaming addicts, in part by exposing them to gaming cues that cause cravings and monitoring their neural responses. These neural changes are basically the same as those seen in other addictive disorders.

So, what do all these brain changes mean? "We focused on how the brain reacts to video game exposure, but these effects do not always translate to real-life changes," says Palaus. As video games are still quite new, the research into their effects is still in its infancy.

For example, we are still working out what aspects of games affect which brain regions and how. "It's likely that video games have both positive (on attention, visual and motor skills) and negative aspects (risk of addiction), and it is essential we embrace this complexity," explains Palaus.

1.6 Video Games Play May Provide Learning, Health, Social Benefits, Review Finds.[27]

Playing video games, including violent shooter games, may boost children's learning, health and social skills, according to a review of research on the positive effects of video game play to be published by the American Psychological Association.

The study comes out as debate continues among psychologists and other health professionals regarding the effects of violent media on youth. An APA task force is conducting a comprehensive review of research on violence in video games and interactive media and will release its findings in 2014.

"Important research has already been conducted for decades on the negative effects of gaming, including addiction, depression and aggression, and we are certainly not suggesting that this should be ignored," said lead author Isabela Granic, PhD, of Radboud University Nijmegen in The Netherlands. "However, to understand the impact of video games on children's and adolescents' development, a more balanced perspective is needed." The article will be published in APA's flagship journal, *American Psychologist*.

While one widely held view maintains playing video games is intellectually lazy, such play actually may strengthen a range of cognitive skills such as spatial navigation, reasoning, memory and perception, according to several studies reviewed in the article. This is particularly true for shooter video games that are often violent, the authors said. A 2013 meta-analysis found that playing shooter video games improved a player's capacity to think about objects in three dimensions, just as well as academic courses to enhance these same skills, according to the study.

"This has critical implications for education and career development, as previous research has established the power of spatial skills for achievement in science, technology, engineering and mathematics," Granic said. This enhanced thinking was not found with playing other types of video games, such as puzzles or role-playing games.

Playing video games may also help children develop problem-solving skills, the authors said. The more adolescents reported playing strategic video games, such as role-playing games, the more they improved in problem solving and school grades the following year, according to a long-term study published in 2013. Children's creativity was also enhanced by playing any kind of video game, including violent games, but not when the children used other forms of technology, such as a computer or cell phone, other research revealed.

Simple games that are easy to access and can be played quickly, such as "Angry Birds," can improve players' moods, promote relaxation and ward off anxiety, the study said. "If playing video games simply makes people happier, this seems to be a fundamental emotional benefit to consider," said Granic.

The authors also highlighted the possibility that video games are effective tools to learn resilience in the face of failure. By learning to cope with ongoing failures in games, the authors suggest that children build emotional resilience they can rely upon in their everyday lives.

Another stereotype the research challenges is the socially isolated gamer. More than 70 percent of gamers play with a friend and millions of people worldwide participate in massive virtual worlds through video games such as "Farmville" and "World of Warcraft," the article noted.

Multiplayer games become virtual social communities, where decisions need to be made quickly about whom to trust or reject and how to lead a group, the authors said. People who play video games, even if they are violent, that encourage cooperation are more likely to be helpful to others while gaming than those who play the same games competitively, a 2011 study found.

The article emphasized that educators are currently redesigning classroom experiences, integrating video games that can shift the way the next generation of teachers and students approach learning. Likewise, physicians have begun to use video games to motivate patients to improve their health, the authors said.

In the video game "Re-Mission," child cancer patients can control a tiny robot that shoots cancer cells, overcomes bacterial infections and manages nausea and other barriers to adhering to treatments. A 2008 international study in 34 medical centers found significantly greater adherence to treatment and cancer-related knowledge among children who played "Re-Mission" compared to children who played a different computer game.

1.7 Neural Basis of Video Gaming: A Systematic Review.[28]

Background: Video gaming is an increasingly popular activity in contemporary society, especially among young people, and video games are increasing in popularity not only as a research tool but also as a field of study. Many studies have focused on the neural and behavioural effects of video games, providing a great deal of video game derived brain correlates in recent decades. There is a great amount of information, obtained through a myriad of methods, providing neural correlates of video games.

Objectives: We aim to understand the relationship between the use of video games and their neural correlates, taking into account the whole variety of cognitive factors that they encompass.

Methods: A systematic review was conducted using standardized search operators that included the presence of video games and neuro-imaging techniques or references to structural or functional brain changes.

Separate categories were made for studies featuring Internet Gaming Disorder and studies focused on the violent content of video games.

Results: A total of 116 articles were considered for the final selection. One hundred provided functional data and 22 measured structural brain changes. One-third of the studies covered video game addiction, and 14% focused on video game related violence.

Conclusions: Despite the innate heterogeneity of the field of study, it has been possible to establish a series of links between the neural and cognitive aspects, particularly regarding attention, cognitive control, visuospatial skills, cognitive workload, and reward processing.

However, many aspects could be improved. The lack of standardization in the different aspects of video game related research, such as the participants' characteristics, the features of each video game genre and the diverse study goals could contribute to discrepancies in many related studies.

1.8 One Hour of Video Gaming Can Increase the Brain's Ability to Focus.[29]

Researchers have demonstrated that just one hour spent playing video games has an effect on the brain. The research team found changes in brain activity and increased performance on tests of visual selective activity in subjects who had spent one hour playing the League of Legends video game. Their results are published in the journal *Frontiers in Human Neuroscience.*

Weiyi Ma, assistant professor of human development and family sciences in the University of Arkansas School of Human Environmental Sciences, collaborated with researchers at the Key Laboratory for NeuroInformation of the Ministry of Education of China, one of the leading research centers for neuroscience in China. Dezhong Yao and Diankun Gong direct the Key Laborator and serve as co-authors of the article.

STUDY DESIGN

Twenty-nine male students at the University of Electronic Science and Technology of China participated in the study. The students were identified as either experts, who had at least two years of experience playing action video games and were ranked in the top seven percent of League of Legends players, and non-experts, who had less than half a year of experience and were ranked in the lowest 11 percent of players.

Before and after playing the video game, the participants' visual selective attention was assessed. Visual selective attention refers to the brain's ability to focus on relevant visual information while suppressing less relevant information. Processing information uses energy, so individuals who excel at visual selective attention—who can narrow their focus and block out distractions—are using their brains more efficiently.

To assess visual selective attention, the researchers briefly showed each subject a square in the canter of a computer screen. Then another square flashed in a different part of the screen and the subject was asked to identify the position of the second square relative to the first. During the course of the experiment, the researchers also monitored brain activity associated with attention using electroencephalography, or EEG.

RESULTS

The researchers observed that in the initial assessment, the expert gamers had more brain activity associated with attention than the non-experts. The experts also scored better on the initial visual selective attention assessment.

After the hour long video game session, both the experts and non-experts had improved visual selective attention, and the two groups received similar scores on the post-game assessment. The non-experts also showed changes in brain activity, according to the EEG data. After the gaming session, their brain activity was similar to that of the experts.

1.9 Video gaming may be associated with better cognitive performance in children.[30]

A study of nearly 2,000 children found that those who reported playing video games for three hours per day or more performed better on cognitive skills tests involving impulse control and working memory compared to children who had never played video games. Published today in *JAMA Network Open*, this study analysed data from the ongoing Adolescent Brain Cognitive Development (ABCD) Study(link is external), which is supported by the National Institute on Drug Abuse (NIDA) and other entities of the National Institutes of Health.

"This study adds to our growing understanding of the associations between playing video games and brain development," said NIDA Director Nora Volkow, M.D.

"Numerous studies have linked video gaming to behaviour and mental health problems. This study suggests that there may also be cognitive benefits associated with this popular pastime, which are worthy of further investigation."

Although a number of studies have investigated the relationship between video gaming and cognitive behaviour, the neurobiological mechanisms underlying the associations are not well understood. Only a handful of neuroimaging studies have addressed this topic, and the sample sizes for those studies have been small, with fewer than 80 participants.

To address this research gap, scientists at the University of Vermont, Burlington, analysed data obtained when children entered the ABCD Study at ages 9 and 10 years old. The research team examined survey, cognitive, and brain imaging data from nearly 2,000 participants from within the bigger study cohort. They separated these children into two groups, those who reported playing no video games at all and those who reported playing video games for three hours per day or more.

This threshold was selected as it exceeds the American Academy of Pediatrics screen time guidelines(link is external), which recommend that videogaming time be limited to one to two hours per day for older children. For each group, the investigators evaluated the children's performance on two tasks that reflected their ability to control impulsive behaviour and to memorize information, as well as the children's brain activity while performing the tasks.

The researchers found that the children who reported playing video games for three or more hours per day were faster and more accurate on both cognitive tasks than those who never played. They also observed that the differences in cognitive function observed between the two groups was accompanied by differences in brain activity.

Functional MRI brain imaging analyses found that children who played video games for three or more hours per day showed higher brain activity in regions of the brain associated with attention and memory than did those who never played. At the same time, those children who played at least three hours of videogames per day showed more brain activity in frontal brain regions that are associated with more cognitively demanding tasks and less brain activity in brain regions related to vision.

The researchers think these patterns may stem from practicing tasks related to impulse control and memory while playing videogames, which can be cognitively demanding, and that these changes may lead to improved performance on related tasks.

Furthermore, the comparatively low activity in visual areas among children who reported playing video games may reflect that this area of the brain may become more efficient at visual processing as a result of repeated practice through video games.

While prior studies have reported associations between video gaming and increases in depression, violence, and aggressive behaviour, this study did not find that to be the case. Though children who reported playing video games for three or more hours per day did tend to report higher mental health and behavioural issues compared to children who played no video games, the researchers found that this association was not statistically significant, meaning that the authors could not rule out whether this trend reflected a true association or chance. They note that this will be an important measure to continue to track and understand as the children mature.

Further, the researchers stress that this cross-sectional study does not allow for cause-and-effect analyses, and that it could be that children who are good at these types of cognitive tasks may choose to play video games. The authors also emphasize that their findings do not mean that children should spend unlimited time on their computers, mobile phones, or TVs, and that the outcomes likely depend largely on the specific activities children engage in. For instance, they hypothesize that the specific genre of video games, such as action-adventure, puzzle solving, sports, or shooting games, may have different effects for neurocognitive development, and this level of specificity on the type of video game played was not assessed by the study.

"While we cannot say whether playing video games regularly caused superior neurocognitive performance, it is an encouraging finding, and one that we must continue to investigate in these children as they transition into adolescence and young adulthood," said Bader Chaarani, Ph.D., assistant professor of psychiatry at the University of Vermont and the lead author on the study.

"Many parents today are concerned about the effects of video games on their children's health and development, and as these games continue to proliferate among young people, it is crucial that we better understand both the positive and negative impact that such games may have."

Through the ABCD Study, researchers will be able to conduct similar analyses for the same children over time into early adulthood, to see if changes in video gaming behaviour are linked to changes in cognitive skills, brain activity, behaviour, and mental health. The longitudinal study design and comprehensive data set will also enable them to better account for various other factors in the children's families and environment that may influence their cognitive and behavioural development, such as exercise, sleep quality, and other influences.

The ABCD Study, the largest of its kind in the United States, is tracking nearly 12,000 youth as they grow into young adults. Investigators regularly measure participants' brain structure and activity using magnetic resonance imaging (MRI) and collect psychological, environmental, and cognitive information, as well as biological samples. The goal of the study is to understand the factors that influence brain, cognitive, and social-emotional development, to inform the development of interventions to enhance a young person's life trajectory.

1.10 Reference:

1. What is Industry 4.0? Everything you need to know about Industry 4.0's Impact on Education. Mentalup . *15 April 2021*

https://www.mentalup.co/blog/industry-4-and-its-impact-on-education

2. Nur Maisarah Ismaizam, Siti Fatimah Abd. Rahman, Sharifah Nurcamelia Syed Mohamad Ahmad, Nur Intan Irdiana Mohd Nazri, Nur Adibah Alya Idris, Nurul Ayreen Ali, Nur Fiza Batrisyia Mohamad Rafi, Syamsul Nor Azlan Mohamad, Azwin Arif Abdul Rahim, Khadijah Khalilah Abdul Rashid, Abdulmajid Mohammed Abdulwahab Aldaba Kulliyyah of Education, International Islamic University Malaysia (IIUM), Malaysia, 8Curriculum Affairs Unit, Universiti Teknologi MARA (UiTM), Malaysia, 9Pusat Bahasa Moden, Universiti Malaysia Pahang (UMP), Malaysia Corresponding author: sfarahman@iium.edu.my. An Integration of Game-based Learning in a Classroom: An Overview (2016 - 2021). International Journal of Academic Research in Progressive Education and Development Vol. 1 1 , No. 1, 2022, E-ISSN: 2226-6348 © 2022 HRMARS. https://hrmars.com/papers_submitted/12347/an-integration-of-

game-based-learning-in-a-classroom-an-overview-2016-2021.pdf

3. Saad, N. M., Baharuddin, J., & Ismail, S. N. (2016). Hubungan Antara Tahap kompetensi fungsional Guru Dengan ... Retrieved January 21, 2022. http://repo.uum.edu.my/23107/1/ICECRS%2C%201%20%282016%29%20199-208.pdf

4. Zhang, F. (2018). The application of the game-based approach in primary school English teaching. Proceedings of the 2nd International Conference on Economics and Management, Education, Humanities and Social Sciences (EMEHSS 2018), 595–596. https://doi.org/10.2991/emehss18.2018.120

5. Jenni Deveau,[1] Susanne M. Jaeggi,[2,3] Victor Zordan,[4] Calvin Phung,[4] and Aaron R. Seitz[1,*] How to build better memory training games Front Syst Neurosci. Published online 2015 Jan 9. doi: 10.3389/fnsys.2014.00243

6. Green, C. S., and Bavelier, D. (2003). Action video game modifies visual selective attention. *Nature* 423, 534–537. doi: 10.1038/nature01647 .

7. Basak, C., Boot, W. R., Voss, M. W., and Kramer, A. F. (2008). Can training in a real-time strategy video game attenuate

cognitive decline in older adults? *Psychol. Aging* 23, 765–777. doi: 10.1037/a0013494

8. Shute, V. J., Ventura, M., and Ke, F. (2015). The power of play: the effects of portal 2 and lumosity on cognitive and noncognitive skills. *Comput. Educ.* 80, 58–67. doi: 10.1016/j.compedu.2014.08.013

9. Franceschini, S., Gori, S., Ruffino, M., Viola, S., Molteni, M., and Facoetti, A. (2013). Action video games make dyslexic children read better. *Curr. Biol.* 23, 462–466. doi: 10.1016/j.cub.2013.01.044

10. Goldin, A. P., Hermida, M. J., Shalom, D. E., Elias Costa, M., Lopez-Rosenfeld, M., Segretin, M. S., et al. (2014). Far transfer to language and math of a short software-based gaming intervention. *Proc. Natl. Acad. Sci. U S A* 111, 6443–6448. doi: 10.1073/pnas.1320217111

11. Hawkins, G. E., Rae, B., Nesbitt, K. V., and Brown, S. D. (2013). Gamelike features might not improve data. *Behav. Res. Methods* 45, 301–318. doi: 10.3758/s13428-012-0264-3

12. Katz, B., Jaeggi, S., Buschkuehl, M., Stegman, A., and Shah, P. (2014). Differential effect of motivational features on training improvements in school-based cognitive training. *Front. Hum. Neurosci.* 8:242. doi: 10.3389/fnhum.2014.00242

13. Rabin, S. (2005). *Introduction to Game Development.* Rockland, MA: Charles River Media.

14 Salen, K., and Zimmerman, E. (2004). *Rules of Play: Game Design Fundamentals.* (MIT Press).

15. Gee, J. P. (2007). *What Video Games have to Teach us about Learning and Literacy: Revised and Updated Edition.* New York, NY: Macmillan.

16. Ericsson, K. A., Krampe, R. T., and Tesch-Römer, C. (1993). The role of deliberate practice in the acquisition of expert performance. *Psychol. Rev.* 100, 363–406. doi: 10.1037//0033-295x.100.3.363

17. Seitz, A. R., Kim, D., and Watanabe, T. (2009). Rewards evoke learning of unconsciously processed visual stimuli in adult humans. *Neuron* 61, 700–707. doi: 10.1016/j.neuron.2009.01.016

18. Adams, E. (2009). *Fundamentals of Game Design (2nd Edn.).* Berkeley, CA: New Riders.Ahissar, M., and Hochstein, S. (1993). Attentional control of early perceptual learning. *Proc. Natl. Acad. Sci. U S A* 90, 5718–5722. doi: 10.1073/pnas.90.12.5718

19. Shibata, K., Yamagishi, N., Ishii, S., and Kawato, M. (2009). Boosting perceptual learning by fake feedback. *Vision Res.* 49, 2574–2585. doi: 10.1016/j.visres.2009.06.009

20. Ahissar, M., and Hochstein, S. (1997). Task difficulty and the specificity of perceptual learning. Nature 387, 401–406. doi: 10.1038/387401a0

21. Salen, K., and Zimmerman, E. (2004). Rules of Play: Game Design Fundamentals. (MIT Press).

22. Ericsson, K. A., Krampe, R. T., and Tesch-Römer, C. (1993). The role of deliberate practice in the acquisition of expert performance. Psychol. Rev. 100, 363–406. doi: 10.1037//0033-295x.100.3.363

23. Walsh, R. (2011). Lifestyle and mental health. Am. Psychol. 66, 579–592. doi: 10.1037/a0021769

24. Sigman, M., Pñna, M., Goldin, A. P., and Ribeiro, S. (2014). Neuroscience and education: prime time to build the bridge. Nat. Neurosci. 17, 497–502. doi: 10.1038/nn.3672

25. Christine Schulz . The Best Mental Games and Exercises to Improve Memory. https://blog.totalbrain.com/the-best-mental-games-and-exercises-to-improve-memory Jun 14, 2021

26. Video games can change your brain Frontiers https://www.sciencedaily.com/releases/2017/06/170622103824.htm **June 22, 2017**

27. Video Games Play May Provide Learning, Health, Social Benefits, Review Finds. 2013 https://www.apa.org/news/press/releases/2013/11/video-games

28. Marc Palaus[1], Elena M Marron[1], Raquel Viejo-Sobera[1,2], Diego Redolar-Ripoll[1]Neural Basis of Video Gaming: A Systematic Review. PMCID: PMC5438999. DOI: 10.3389/fnhum.2017.00248 2017 May 22.

29. Weiyi Ma, Camilla Shumaker. One Hour of Video Gaming Can Increase the Brain's Ability to Focus. Feb. 14, 2018. University of Arkansas https://news.uark.edu/articles/40981/one-hour-of-video-gaming-can-increase-the-brain-s-ability-to-focus

30. Video gaming may be associated with better cognitive performance in children https://www.nih.gov/news-events/news-releases/video-gaming-may-be-associated-better-cognitive-performance-children#:~:text=A%20study%20of%20nearly%202%2C000,had%20never%20played%20video%20games. October 24, 2022

Chapter 2: Game Based Learning

Peter Chew Triangle Diagram Rules

Peter Chew http://dx.doi.org/10.2139/ssrn.3847144

Abstract.

Background: The preprint study of Peter Chew's triangle diagram and application [Chew, Peter, 2021] mentioned that the purpose of Peter Chew's triangle diagram is to clearly illustrate the topic solution of the triangle and provide a complete design for the knowledge of the AI era. Peter Chew's triangle diagram will propose a better single rule that allows us to directly, more easily, and more accurately solve any solution of the triangle problem.

Therefore, it is important to create a game that can help learn all the Peter Chew Triangle Diagram rules easily and fun. In addition, according to the preprint study, "The memorization techniques of Peter Chew's Rules [Chew, Peter, 2021]. The memorization techniques of Peter Chew Rule has also been created. Therefore, the goal of the game can also be used to test the efficiency of Peter Chew Rule memorization techniques.

Method: The home page for the game. If the player wants to take a look at the rules before the game, he can click to photo reading. If player want to start the game, just press the PCET program logo. The time given to each player is 30 seconds. Players need to press the PCET logo that moves randomly. If the player successfully presses the PCET logo, a sentence will be displayed and the correct and incorrect logos will be displayed below.

When the right and wrong logos flash, the player can start to press the logo. If the statement is correct, the player needs to press the correct logo, and then the player will receive a score of 5 points, plus a 5-second timer, and the problem will disappear.

If the player answers incorrectly or the timer = 0, high score ranking page will come out. When player press key Enter, the game will restart. When the player reaches 100 points, a page will be displayed saying that you won the game.

Result: Some students have been tested by playing games. First of all, without teaching memorization techniques of Peter Chew Rules, no student can win the game.After that, the students were asked to teach them the memorization techniques of Peter Chew Rules for win the game. After teaching the memorization techniques of Peter Chew Rules, they won the game. This proves that memorization techniques of Peter Chew Rules are effective.

Workshop Game Base Learning for memorize maths rule has been held at Conference on Science and Mathematics Education [Cosmed 2019] by Southeast Asian Ministers of Education Organization, Regional Centre for Education in Science and Mathematics (SEAMEO RECSAM) . Videos related to this can be found on this link https://youtu.be/yV5soDsi7TU .

Conclusion : The traditional teaching method is that the teacher asks students to learn the rules of mathematics, rather than some students want to learn the rules of mathematics, which makes it difficult for students to remember the rules of mathematics.

Through the game-based learning method, because students want to win the game, students will ask the teacher to teach them the rules, which makes it easy for students to learn the rules. In addition, students need to concentrate on thinking about the rules when playing games, which allows them to remember for a longer period of time. In addition, From Healthline, playing memory games can help you get better at concentrating. Brain training games can also help you develop your working and short-term memory, as well as your processing and problem-solving skills.

Keyword; Peter chew triangle diagram, solution of triangle.

2.1 Background

The preprint study , Peter Chew Triangle Diagram and Application [Chew, Peter, 2021] mention objective of Peter Chew Triangle Diagram is to clearly illustrate the topic solution of triangle and provide a complete design for the knowledge of AI age.

Peter Chew's triangle diagram will suggest a better single rule that allows us to solve any problem of topic solution of triangle problems directly, more easily and more accurately. Therefore, it is important to create a game that can help learn all of Peter Chew Triangle Diagram rules easily and fun.

In addition, The preprint study, Memorization Techniques for Peter Chew Rule [Chew, Peter, 2021] is having a Memorization Techniques for Peter Chew Rule According to the study , publish the Journal of Physics: Conference Series, Peter Chew Rule for Solution of Triangle [Chew, Peter, 2019] .

Objectives Peter Chew rule: is to let upcoming generation solve same problem can solve directly, more easy, more accuracy compare what`s now solution. With sine rule, cosine rule and Peter Chew rule, all problem at solution of triangle can solve directly more easy, more accuracy. Peter Chew rule are complement rule of topic solution of triangle.

The International Journal of Game-Based Learning (IJGBL)(https://www.igiglobal.com/journal/international-journal-game-based-learning/41019) is devoted to the theoretical and empirical understanding of game-based learning. To achieve this aim, the journal publishes theoretical manuscripts, empirical studies, and literature reviews. The journal publishes this multidisciplinary research from fields that explore the cognitive and psychological aspects that underpin successful educational video games.

The target audience of the journal is composed of professionals and researchers working in the fields of educational games development, e-learning, technology-enhanced education, multimedia, educational psychology, and information technology. IJGBL promotes an in-depth understanding of the multiple factors and challenges inherent to the design and integration of Game-Based Learning environments.

Game-Based Learning Studies in Education Journals: A Systematic Review of Recent Trends [Uğur Bakan et all , 2018] 119 Abstract: This study investigates descriptively the use of game-based applications in learning and teaching environments.

Each empirical finding was categorized according to paper title, year of publication, number of citations in the Web of Science (WoS) database in SCI, SSCI and AHCI, learning theory, learning principle, game genre, game design elements, learning outcomes and research skills samples, and learning domains.

A total of 190 original game related research articles were selected as sources in six peer-reviewed journals with a 12-year period from 2005 to 2017. The research found that game studies, as well as the cognitive understanding and application-level knowledge of the field are more effective in learning and in student achievements in terms of retention.

A study , publish by PMC, Does Video Gaming Have Impacts on the Brain: Evidence from a Systematic Review [Denilson Brilliant T. et all, 2019] The systematic review has several conclusions related to beneficial effects of noncognitive-based video games.

First, noncognitive-based video gaming can be used in all age categories as a means to improve the brain. However, effects on children remain unclear. Second, noncognitive-based video gaming affects both structural and functional aspects of the brain. Third, video gaming effects were observed after a minimum of

16 h of training. Fourth, some methodology criteria must be improved for better methodological quality.

In conclusion, acute video gaming of a minimum of 16 h is beneficial for brain function and structure. However, video gaming effects on the brain area vary depending on the video game type.

A study , publish by PMC, How to build better memory training games [Jenni Deveau, et all, 2015] . The study suggest that more integrative approaches will lead to better learning outcomes. We suggest that the general mechanisms that promote PL are shared across brain regions and will also promote WM. Furthermore, there is enough known about the aspects of conventional video games that lead to positive learning outcomes that these principles can be applied to achieve more effective WM training.

Additionally, there are other principles, that were beyond the scope of the present review, such as deliberate practice (Ericsson et al., 1993), and many aspects of healthy lifestyles (Walsh, 2011; Sigman et al., 2014) that also promote cognitive fitness. Integrating these approaches with good design could lead to a more comprehensive impact on WM function that might ultimately transfer to real-world conditions.

Many game design criteria mirror components found to improve learning from the PL literature, and literature on deliberate practice (Ericsson et al., 1993). For example, consistent reinforcement to training stimuli (Seitz and Watanabe, 2009) maps directly to consistent player feedback, a key part of player-centric interface design (Adams, 2009).

Adams says (of players) "most critically, they need information about whether their efforts are succeeding or failing, taking them closer to victory or closer to defeat".

Likewise, motivating tasks (Shibata et al., 2009) and ensuring subjects are confident of their performance (Ahissar and Hochstein, 1997) are consistent with good game-design principles such as establishing clear goals (Salen and Zimmerman, 2004) and balancing games challenges (related to the adaptive approaches used in PL and WM training) to match player performance (Adams, 2009). Applying video-game techniques purposefully into WM training can inject the cognitive benefits found from off-the-shelf video games into principled cognitive training, while also being fun to play.

A study , publish by PMC, Neural Basis of Video Gaming: A Systematic Review [Marc Palaus, et all, 2017] . The results suggest playing video games may lead to various changes in the brain, including increased attention and focus.

A study , Rapid Improvement in Visual Selective Attention Related to Action Video Gaming Experience[Nan Qiu et all , 2018] , the study revealed that improvement in VSA is observable after a 1 h AVG training session. Furthermore, both long-term and brief AVG experience influenced cognitive processes related to attentional selection and control process.

VSA refers to your ability to concentrate on a specific task while ignoring distractions around you. This study was limited by its small size, so these findings aren't conclusive. The study also didn't determine how long this increase in VSA lasted.

2.2 . Method .

i) The home page for the game.

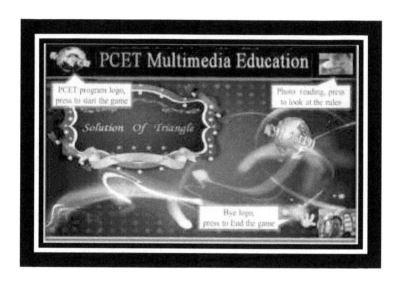

ii) The rules page .

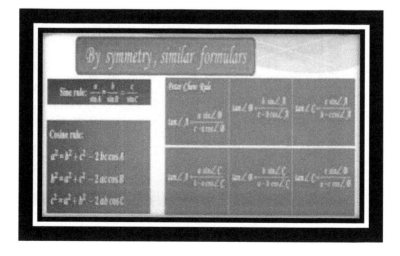

iii) Game start.

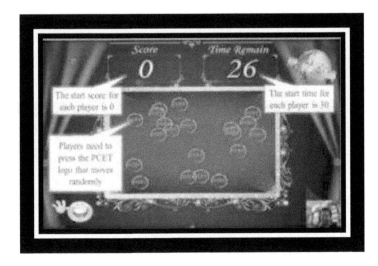

iv) If the player successfully presses the PCET logo, a sentence will be displayed and the correct and incorrect logos will be displayed below.

v) If the player answers incorrectly or the timer = 0, high score page will come out.

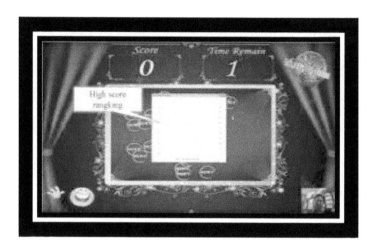

vi) When the player reaches 100 points, a page will be displayed saying that you won the game.

2.3 Result.

Some students have been tested by playing games. First of all, without teaching memorization techniques of Peter Chew Rules, no student can win the game. After that, the students were asked to teach them the memorization techniques of Peter Chew Rules for win the game. After teaching the memorization techniques of Peter Chew Rules, they won the game. This proves that memorization techniques of Peter Chew Rules are effective.

i)Some students have been tested by playing games.

Workshop Game Base Learning for memorize maths rule has been held at Conference on Science and Mathematics Education [COSmed 2019] by Southeast Asian Ministers of Education Organization, Regional Centre for Education in Science and Mathematics (SEAMEO RECSAM) . Videos related to this can be found on this link https://youtu.be/yV5soDsi7TU .

ii) b) i) Before teaching memorization techniques of Peter Chew Rules, no one can win the game .

ii) b) ii) Teaching memorization techniques of Peter Chew Rules.

ii) b) iii) After teaching memorization techniques of Peter Chew Rules. , the first winner .

iii) Workshop Game Based Learning for memorize maths rule held at the Center Prai District Education Office, Penang , Malaysia.

In the year 2022, the Seberang Prai Tengah District Education Office in the picturesque region of Penang hosted a captivating and immersive workshop that revolved around the innovative concept of game-based learning. This workshop, designed to foster engaging and effective teaching strategies, became a memorable event for both participants and organizers alike.

Within the workshop, seven Additional Mathematics Teacher emerged as winners, showcasing their exceptional grasp of the principles and practices of game-based learning. Their achievements not only served as a testament to their dedication but also inspired others to explore new avenues of teaching and learning.

The workshop boasted an impressive attendance of 25 Additional Mathematics Teacher. The discussions and activities that unfolded during the workshop were characterized by a deep focus on Additional Mathematics, which allowed for the exchange of advanced insights and pedagogical strategies.

What set the 2022 workshop apart from its predecessor, the COSmed Workshop of 2019, was the composition of its participants. In stark contrast to the diverse group present in 2019, the 2022 workshop brought together Additional Mathematics Teacher . This homogeneity enabled a more concentrated and tailored approach to teaching and learning, resulting in significantly improved outcomes.

The success of the 2022 workshop serves as a shining example of how targeted professional development can yield substantial benefits. By focusing on the specific needs and interests of the participants, this workshop not only enhanced the skills of the educators but also contributed to the overall advancement of educational practices in the region.

2.4 Conclusion :

The traditional teaching method is that the teacher asks the students to memorise the rules of mathematics, but some students are too lazy to memorize the rules of mathematics, because it is not what they want. If teachers force students to memorise math rules, it may cause students to develop a phobia of mathematics.

Through the game-based learning method, because students want to win the game, students will ask the teacher to teach them the rules, which makes it easy for students to remember the rules, because remembering the rules has become what they want.

In addition, students need to concentrate on thinking about the rules while playing the game, which allows them to memorize the math rules for a longer time. This will make teaching and learning mathematics more effective.

Playing memory games can help students focus better. Brain training games can also help students develop their working memory and short-term memory, and enhance students' ability to process and solve problems.

2.5. References

1.Adams E. (2009). Fundamentals of Game Design (2nd Edn.). Berkeley, CA: New Riders. Ahissar M., Hochstein S. (1997). Task difficulty and the specificity of perceptual learning. Nature 387, 401–406. 10.1038/387401a0 1997 May 22;387(6631):401-6. doi: 10.1038/387401a0.

2.Chew, Peter, Peter Chew Rule for Solution of Triangle (2019). 2019 the 8th International Conference on Engineering Mathematics and Physics, Journal of Physics: Conference Series1411 (2019) 012009, IOP Publishing, doi:10.1088/1742-6596/1411/1/012009, Available at SSRN: https://ssrn.com/abstract=3843433

3.Chew, Peter, Memorization Techniques for Peter Chew Rule (March 5, 2021). Available at SSRN: https://ssrn.com/abstract=3798502 or http://dx.doi.org/10.2139/ssrn.3798502

4.Chew, Peter, Peter Chew Triangle Diagram and Application (March 5, 2021). Available at SSRN: https://ssrn.com/abstract=3798488 or http://dx.doi.org/10.2139/ssrn.3798488

5.Denilson Brilliant T., Rui Nouchi, and Ryuta Kawashima Does Video Gaming Have Impacts on the Brain: Evidence from a Systematic Review. PMC. Published online 2019 Sep 25. doi: 10.3390/brainsci9100251

6.Ericsson K. A., Krampe R. T., Tesch-Römer C. (1993). The role of deliberate practice in the acquisition of expert performance. Psychol. Rev. 100, 363–406 10.1037//0033-95x.100.3.363

7. Jenni Deveau, 1 Susanne M. Jaeggi, 2,3 Victor Zordan, 4 Calvin Phung, 4 and Aaron R. Seitz1,* How to build better memory training games. PMC .Published online 2015 Jan 9. doi: 10.3389/fnsys.2014.00243

8.Marc Palaus, 1,* Elena M. Marron, 1 Raquel Viejo-Sobera, 1,2 and Diego Redolar-RipollNeural Basis of Video Gaming: A Systematic Review. PMC. Published online 2017 May 22. doi: 10.3389/fnhum.2017.00248

9.Nan Qiu1,2 , Weiyi Ma3 , Xin Fan 1,2 , Youjin Zhang1,2 , Yi Li 1,2 , Yuening Yan , Zhongliang Zhou 1,2 , Fali Li 1,2 , Diankun Gong and Dezhong Yao Rapid Improvement in Visual Selective Attention Related to Action Video Gaming Experience.

Front. Hum. Neurosci., 13 February 2018 |
https://doi.org/10.3389/fnhum.2018.000471

10.Salen K., Zimmerman E. (2004). Rules of Play: Game Design Fundamentals. (MIT Press;). Seitz A. R., Watanabe T. (2009). The phenomenon of task-irrelevant perceptual learning. Vision Res. 49, 2604–2610. Published 2009 Aug 7. doi: 10.1016/j.visres.2009.08.003

11.Shibata K., Yamagishi N., Ishii S., Kawato M. (2009). Boosting perceptual learning by fake feedback. Vision Res. 49, 2574–2585. 2009 Oct;49(21):2574- 85. doi: 10.1016/j.visres.2009.06.009.

12.Ugur Bakan İzmir , Ufuk Bakan İzmir . Game-Based Learning Studies in Education Journals: A Systematic Review of Recent Trends. . DOI:10.19052/ap.5245

13.Walsh R. (2011). Lifestyle and mental health. Am. Psychol. 66, 579–592. 10.1037/a0021769

9 798223 215813